ROCKS

by Meg Gaertner

Cody Koala

An Imprint of Pop!
popbooksonline.com

abdobooks.com
Published by Pop!, a division of ABDO, PO Box 398166, Minneapolis, Minnesota 55439. Copyright © 2020 by POP, LLC. International copyrights reserved in all countries. No part of this book may be reproduced in any form without written permission from the publisher. Pop!™ is a trademark and logo of POP, LLC.

Printed in the United States of America, North Mankato, Minnesota

052019
092019

THIS BOOK CONTAINS RECYCLED MATERIALS

Cover Photo: Shutterstock Images
Interior Photos: Shutterstock Images, 1, 5 (top), 6, 9, 10, 13, 14–15, 17, 18–19; iStockphoto, 5 (bottom left), 5 (bottom right), 21

Editor: Connor Stratton
Series Designer: Sarah Taplin

Library of Congress Control Number: 2018964778

Publisher's Cataloging-in-Publication Data
Names: Gaertner, Meg, author.
Title: Rocks / by Meg Gaertner.
Description: Minneapolis, Minnesota : Pop!, 2020 | Series: Science all around | Includes online resources and index.
Identifiers: ISBN 9781532163609 (lib. bdg.) | ISBN 9781532165047 (ebook)
Subjects: LCSH: Rocks--Juvenile literature. | Geology--Juvenile literature. | Geoscience--Juvenile literature. | Science--Juvenile literature.
Classification: DDC 552--dc23

Hello! My name is

Cody Koala

Pop open this book and you'll find QR codes like this one, loaded with information, so you can learn even more!

Scan this code* and others like it while you read, or visit the website below to make this book pop.

popbooksonline.com/rocks

*Scanning QR codes requires a web-enabled smart device with a QR code reader app and a camera.

Table of Contents

Chapter 1

Igneous Rocks

Rocks are made of more than one **mineral**. Rocks are not living. They are found in nature. There are three types of rocks. They form in different ways.

Watch a video here!

Magma is called lava when it reaches Earth's surface.

One kind of rock is **igneous** rock. These rocks form when **magma** cools and hardens. Magma is rock that has melted from great heat. Magma comes from deep below Earth's surface.

Chapter 2

Sedimentary Rocks

Another kind of rock is **sedimentary** rock. These rocks form from sand, shells, pebbles, and other rocks. All of this **sediment** piles up. It gets buried.

Learn more here!

Over millions of years, the sediment sticks together. It forms sedimentary rock. Sedimentary rock often shows the layers of sediment that formed it.

Sedimentary rocks often contain the remains of long-dead plants and animals.

Chapter 3

Metamorphic Rocks

The last type of rock is **metamorphic** rock. These rocks are made from other rocks. A lot of heat and pressure can turn any rock into a metamorphic rock.

Learn more here!

Metamorphic rocks often have layers that look like ribbons. These layers are

from where the original rock was stretched or folded by heat or pressure.

Chapter 4

The Rock Cycle

Rocks change over millions of years. Weather can break down rocks on Earth's surface. Then, **erosion** carries those pieces away. They become **sediment**.

Complete an activity here!

Slowly, rocks can be pushed under the ground. Deep inside the planet, rocks

can experience great heat and pressure. These rocks become **metamorphic** rocks.

Inside the planet, rocks can melt back into **magma**. Earth pushes that magma to the surface. It cools into **igneous** rock. Later, weather breaks that rock into sediment. The rock cycle continues.

The Rock Cycle

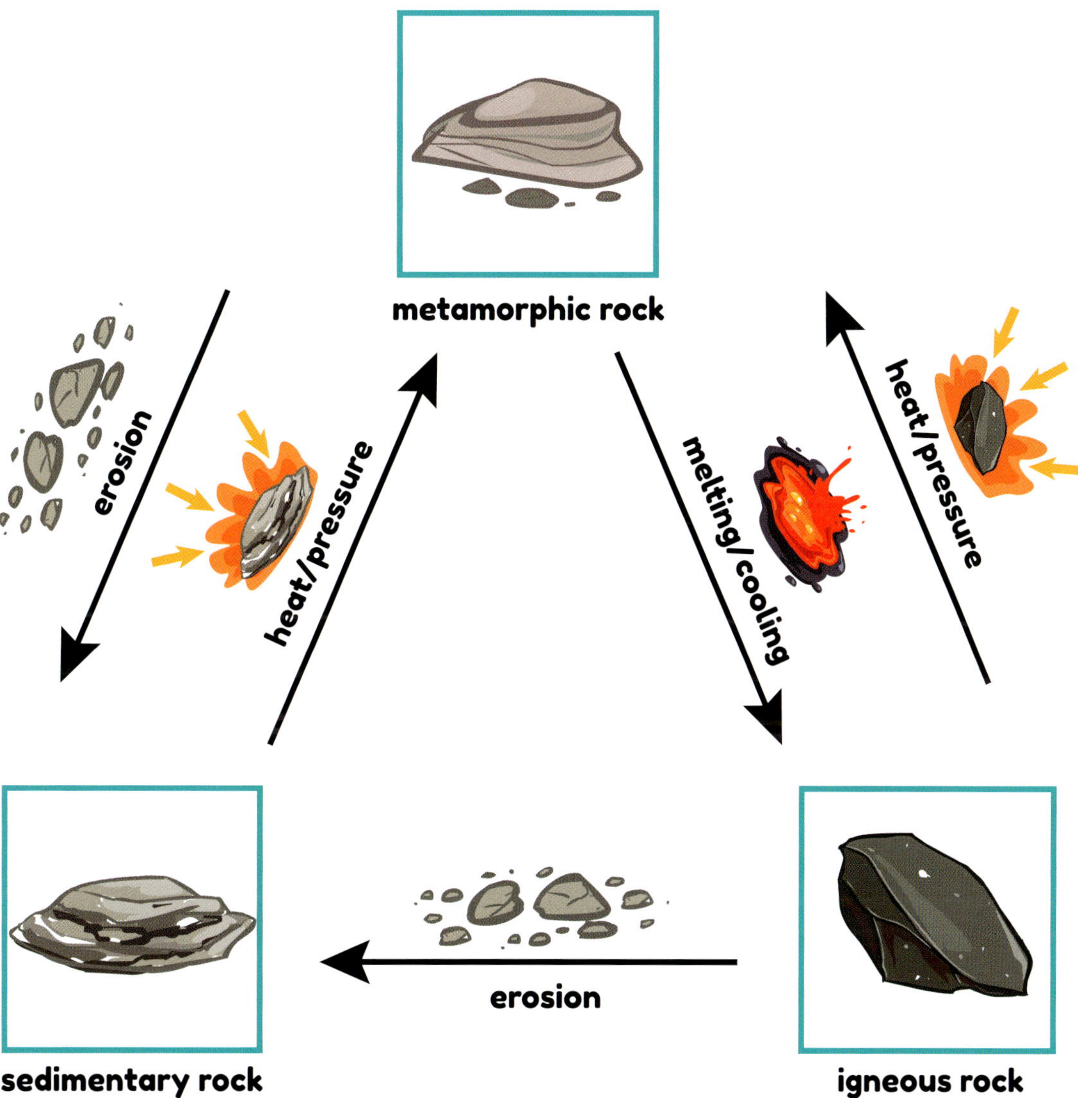

Making Connections

Text-to-Self

Where do you see rocks in your life? What do they look like?

Text-to-Text

Have you read other books about rocks or minerals? What did you learn?

Text-to-World

Rocks change form through the rock cycle. What else changes form over time?

Glossary

erosion – the process of wind or water moving sediment to a new area.

igneous – a type of rock formed when magma cools and hardens.

magma – rock that has melted from intense heat.

metamorphic – a type of rock formed from other rocks that experience great heat or pressure.

mineral – a nonliving solid that forms naturally.

sediment – bits of rocks, shells, and other things.

sedimentary – a type of rock formed when layers of sediment become solid over time.

Index

Online Resources

popbooksonline.com

Thanks for reading this Cody Koala book!

Scan this code* and others like it in this book, or visit the website below to make this book pop!

popbooksonline.com/rocks

*Scanning QR codes requires a web-enabled smart device with a QR code reader app and a camera.